THE SECRETS OF THE ENERGIES AT THE MEGALITHS

IN

CARNAC & BRITTANY

Measured with the Lecher Antenna

Sequel

Written by Dame Anne-Marie Delmotte

Copyright2020 Anne-Marie Delmotte

Disclaimer: The author of this book does not dispense medical advice or prescribe the use of any technique as a form of treatment for physical or medical problems without the advice of a physician, either directly or indirectly. The intent of the author is only to offer information of a general nature to help you in your quest for well-being. In the event you use any of the information in this book for yourself, which is your constitutional right, the author and publisher assume no responsibility for your actions.

ISBN: 9789082802696
EAN: 978-90-82802-69-6
NUR2: 402
AWS: W

ABOUT THE AUTHOR

Dame Anne-Marie Delmotte has an Associate's Degree in Clinical Chemistry and works as a scientist at the Belgian Government.

In 2018, she received the title of Knight in the Leopold Order from the King of Belgium for her work. The work she does for the government is completely different from her own personal energy research project.

Because she is hypersensitive to all kinds of energies and wanted to objectify what she was sensing, she searched for an instrument that is able to measure and determine these energies objectively and scientifically. In Belgium, in 2008, she was introduced to the Lecher antenna by the non-profit organisation CEREB - Belgium from whom she acquired her so long for searched instrument. She attended several Lecher antenna training courses in geobiology and bio-energy measurements. She set off to do measurements at places that are known for their special energies like the Chartres Cathedral near Paris and several megalithic sites in Ireland and France. After the miraculous recovery of her mother, who was dying, after a balancing of her energies with the Lecher antenna she decided to learn even more about bio-energy balancings. She writes about her elaborate research in her books and has developed several e-learning courses and a Practical Guide in which she teaches others how to work with the Lecher antenna.

Anne-Marie is currently on a several year career break to make this precious little instrument that is the Lecher antenna and its many uses much better known than it is today.

She wants to make the knowledge available and much more affordable to all by means of her Practical Guide, books, online training and actual training courses. She herself had to set up a crowdfunding at the time for her own training.

She is constantly looking for low budget solutions to balance a person's bio-energies, to reduce the influence of electromagnetic radiation and to clear negative energies, or what some call geopathic stress that can be hazardous to health and cause illness, from a person's home place.

Her very elaborate research of the Megalithic Sites in mainly Ireland and France have made her understand that the Ancient Builders were already clearing this very geopathic stress thousands of years ago by means of certain stones for better livestock farming, crops and yields. She is able to recreate these very same energies as the Ancients in home places allowing for clearing of the negative energies with very low budget solutions. This way much more people can be helped and her courses teaching this to others makes this knowledge available for even more, of which some unfortunately wait till they are already very sick before seeking aid at an energetic level, to be helped. But the more people know about geopathic stress and bio-energy balancings, the more people might make the link to their energetic health and/or home place as being a possible cause of their "ill" feeling. Anne-Marie's books and work are important into creating more awareness about this. A very noble mission.

I wish for her to be able to keep up this honourable work for many years to come. The world will be a better place for it. Thank you, Anne-Marie, for temporarily stepping down from your important government job serving your country to make this possible.

Anne-Marie also holds following qualifications: USUI Reiki, Shamanic Reiki and Reiki Crystal Master/Teacher/Practitioner Degree as well as a Reiki Space Clearing Practitioner Degree.

CONTENTS

1. INTRODUCTION: HOW ARE THE ENERGIES MEASURED

All energies for this book were measured with a Lecher antenna, a scientific instrument that can send and receive energy and which has a movable cursor and a ruler to select the energy its user wants to measure or send. It measures by means of resonance between what is looked for and what is being found.

I wrote a "Practical Guide for Dowsing with the Lecher Antenna – Elaborate Basic Training Course in Geobiology and Bio-energy" and made e-learning courses in English and in French if should you wish to learn more about the instrument and make your own measurements.

Picture: my personal Lecher antenna

The Lecher antenna I use has a magnetised rod in one of its arms which allows its user to measure in positive and negative mode. This way of measuring has turned out to be extremely useful for the research I write about in this book. This research is strictly my own and as far as I know never researched before. It has therefore not been easy for me to draw firm conclusions, but I am convinced that what you will read in this book is valid information based on many scientific measurements in Brittany, of which some are more than a year apart, proof that the readings are stable and reproducible.

To better understand my findings in this book I would like to add the following information about the energies as such, and measuring in positive and negative mode.

I was taught that an energy signal consists of two spirals.

One spiral is turning clockwise and is positive or beneficial to most life on earth. I write most life on earth, as certain animals and plants are known to thrive on negative energies but these are quite rare.

The other spiral is turning anticlockwise and is negative or harmful to most life on earth. These spiral signals can sometimes be measured with the Lecher antenna depending on whether their signals are strong enough. Most times it is only possible to measure the one energy signal in positive mode and the one signal in negative mode and not the spiral signals.

You might think that these two spirals would neutralise each other, but elaborate research conducted by several scientists and Medical Doctors leads to conclude otherwise, certain of these similar energies are quite harmful to our health and well-being when emanating from the earth at our home place.

I hope you will enjoy being with me on my journey in Brittany where I further ascend the veil on the knowledge of the Ancient builders of these magnificent places I visit. It is recommendable to read my earlier book about my research in Brittany that is called "Signpost to the Holy Grail? Infinite Energy with Infinite Possiblities! Dowsed with the Lecher Antenna in Carnac and Brittany France" as well. It will allow you to even better understand the new discoveries I made during my journey that I describe in this book.

In order for you to experience being with me on this journey, and enjoy it as much as I did, I have added quite a lot of photographs which I took during my visits.

2. THE SPECIAL ENERGIES AT THE VARIOUS MEGALITHIC SITES

Firstly I will explain about two main major types of energy configuration which I have discovered at the Megalithic Sites. This will enable you to be up to date with my research and help you understand the new discoveries I have made. A Megalith is a monument made up of one or more large stones.

1° Energy configuration which brings in many positive beneficial energies of which the interior is free of negative energies. The latter is because a protective shield is in place just outside of the monument. This type of energy configuration creates a wonderful space for creative work, meditation, writing, ... just to be in. This is an energy configuration that I had until now only found in the Carnac region of Brittany at the Quadrilatère of Manio and the Quadrilatère of Crucuno but more recently also at Moyturra and near the Carrowkeel Cairns in Ireland.

Note: *I am able to configure houses, apartments, offices, outdoor spaces, ... into this type of energy configuration shifting out the negative energies, creating a protective shield and bringing in a set of positive, beneficial energies into each room of the house, even the hallway, no matter the angle of the home place and no matter its dimensions by means of certain stones similar to those at Megalithic Sites.*

I only found Megalithic Sites of the same energy configuration after I was already able to make these configurations myself in someone's home which I find quite amazing and puzzling. Recently I have also found a way of absorbing the negative energies which are just outside the protective shield which is a very good thing as this allows to put in the same configuration when a home place is right beside another one (terraced houses, apartments, flats, ...) and where this negative energy would have had an influence on the neighbours' place. Measurements of the energies at Megalithic Sites in Brittany, and in Ireland, have given me the insight that the Ancients were already making similar energy configurations thousands of years ago.

 2° Sites which transform the energies into beneficial energies that have a healing effect on humans, animals and plants. Some of these sites also form a protective shield as in the first configuration which shifts out the negative energies.

Each site has a different energetic quality, although reproducible energy patterns are definitely the case. Every site also has a different "feel" to it which can depend on the person(s) visiting. I would like to leave that for you to discover for yourself. It would be advisable to switch off your phone and take off your watch if it's running on a battery when trying to "feel" the energies present at the Megalithic Sites.

The sites which emanate tachyon energy besides sacred or as some call it spiritual energy and divine energy should, to my knowledge, definitely have healing properties given the nature of this type of energy. David Wagner and Gabriel Cousens, MD who have extensively studied tachyon energy claim that it is the source of all healing. I explain tachyon energy extensively in the glossary at the end of this book.

Please honour and show respect for the Sites and the land. These structures were built for specific reasons beneficial to life on earth. Show them your love and gratitude for serving us all for thousands of years. After reading this book you will have a better insight into why and where these sites were built. Building these sites took an enormous amount of time and effort and a lot of thought was put into where and how these sites had to be built before constructing them. Researching the Megalithic Sites to me has been a journey of discovery upon discovery and sheer amazement at how intelligent and connected to the land and the earth that Neolithic* Man and Woman must have been. There is so much we can learn from these sites.

Words that are explained in the glossary at the end of the book are marked with a *.

The Lecher antenna* can also be used for balancing a person's bio-energies and clearing home places. I once "treated" a remarkable case by doing a bio-balancing and clearing the person's home place by putting in the first energy configuration using local materials found in Ireland. This person knew that she had a problem with certain energies but was so desperate when she first came to me, she wanted to end her life. My interventions with the Lecher antenna* made an enormous difference to this person's life and well-being. About 40 percent of this person's house was built on a stream running to the sea. This running stream, as thoroughly researched by many scientists and Medical Doctors, emanates negative energy which eventually causes ill health. Even more so when a person's bed is located on the running stream. I am so very grateful that the Megalithic Sites allowed me to discover that these negative energies can be worked on and in certain cases, very much so at the different sites in Brittany and in Ireland, transformed into beneficial energies to serve plant, animal and man.

3. ARRIVAL IN CARNAC

I can't believe it. I am back in Carnac for a month and at the same place where I left off last time. I wasn't planning to write but the urge to do so is too great. And I do feel that I have to write because I get the distinct feeling that I will again make new discoveries and by not writing on a regular basis I would not be able to take you, dear reader, with me into the excitement of my new Lecher antenna* journey and adventure. I wonder what Brittany has in store for me this time. My aim of this stay here is to promote the Lecher antenna*, my French online training courses and my French books. But I feel it is going to be way more than that. I'm so excited! I'm back at the same AirBnB I was when I left. Karine, my host, who is so nice and makes sure that I'm well-tended for gave me a warm welcome yesterday, as did Osiris or Poupette, Karine's cat. Poupette is still as acrobatic as when I left her and still chases flies to high up in the trees. She carefully watched me when I was moving in yesterday, a rarity as she prefers to play and chase whatever catches her eye. So, yes, I am right back where I left last time.

Poupette is such a nice cat that I can't but help putting a picture of her in this book.

Picture: Osiris or Poupette

12

This time I left my Irish Megalithic Heaven, Sligo to catch the ferry, the WB Yeats, from Dublin to Cherbourg. It is still early in the year and it was quite windy but this new ferry seems to tackle the waves quite well. The captain of the WB Yeats warned us that the Irish Sea would get rough later in the evening, which it did, I could actually feel myself slipping somewhat in my bunk bed, but the car and I made it in one piece.

4. VISIT OF THE CARNAC ALIGNMENTS

I am going to see the Carnac alignments*, or shall I call them the tachyon fields, this afternoon. It's cloudy and dry today but the weather should be sunny and mild the coming week.

Once at the Carnac alignments*, I went to the Menec alignments* first but I saw that all the fences were closed. I know from last time that not all the alignments* can be visited at the same time as they use some of the sites for their sheep to graze on, and others to ensure the vegetation is not ruined by the multitude of visitors who go there. I went across the road to the "maison des mégalithes" to enquire which parts of the alignments* are open at this time. I was pleasantly surprised to find that there were places open that I couldn't visit the last time I was here to do my research.

I decided to go the Kermario Giants and dolmen* first. The Kermario alignments* consists of 1029 stones in ten rows, about 1300 meters (4300 feet) in length.

Wow, it felt amazing to finally walk between these rows of standing stones and just admire them and to see the dolmen* from all sides. I did not do any measurements with my Lecher antenna* as I was still very tired from travelling here, and I just wanted to take in the

13

energies as such and have a good look at the stones which were very enjoyable.

Here are some pictures of some of the standing stones at the Kermario alignments*.

I also walked around the part of the Kermario alignments* where the menhirs (=word for standing stone in Breton) are the least tall.

Some of the rows seem to converge to a certain point in some part. Oddly enough when I was looking down these rows, the sun came out very briefly and the sun-rays moved along down these rows. A sign? Maybe this converging is something to research.

Picture: converging of rows where the standing stones of the Kermario alignments* are the least tall

I found myself all alone in the part of the Kerlescan alignments*. Megalithic heaven! I'm just looking at some plans of the Kerlescan alignments* and I'm wondering whether the enclosure* at the head of the rows of menhirs was meant to be a rectangular (quadrilatère*) shape rather than a circular shape.

Picture: the Kerlescan alignments*

I called it a day because I still had to get some food and write this. I am now in my house which is on the outskirts of Carnac in the midst of farmland enjoying the peace and quiet.

5. EARLY SPRING

Yesterday started with the most beautiful sunrise which I could see from the house. It was only when I stepped out that I realised it to be quite cold. The wind also picked up during the day and we had some short downpours too. I decided to stay in as I still had lots of office work to do, as usual!

I recently bought a new feature for my websites so I would be able to add a "contact form" to my English and French website. I hadn't been able to get it to work up until now, so it was as good a time as any to get this done. After some research and some hours of trial and error I was about to give up but then I tried another configuration and this seemed to do the trick.

I want to publish a second edition of my first book, "The Lecher Antenna Adventures and Research in Geobiology and Bio-energy" or with e-book title "Ascending the Veil on Secret Energies of Megalithic Sites, Healing Energy and Creating Sacred Space Using Free Magical Stones".

I have conducted a lot of research since I wrote this book. I also worked out my own basic method for bio-energy balancings for which some of the people I balanced, certainly those hypersensitive to energies, told me that they experience this method as "softer", less strong and easier to cope with. It is also somewhat shorter than the methods I was using before and when I ask the Lecher antenna* which method to use, it selects almost always my method. It is also the method I teach in my Practical Guide and online e-learning bio-energy Lecher antenna* course level 1 and 2.

I asked a renowned insurance broker who specialises in alternative therapies, whether my method could be considered to be covered and after handing in a file with detailed information about my method, my background and case studies, they accepted. I am sure my clinical background and the integration of my knowledge of Reiki, anatomy and physiology must have worked in my favour towards their decision. I consider this a major breakthrough and a step closer towards bio-energy balancings with the Lecher antenna* being recognised by health insurance companies as a "treatment".

This said, I want to add my method to my first book and put in some of the cases I have "balanced" using my method, I decided to call "Delmotte Vibrating Energies Résonnantes or DELMOTVIBRES" method, the same name as my trademark.

Today seemed to be a dry but overcast day. Unfortunately, I don't feel so well, I am experiencing some minor pains and am quite fatigued. So I decided to treat myself to a self-balancing. This was good thinking as the fatigue is now gone and this balancing might be a good case for the second edition of my first book. If the weather stays fine I might go for a walk along the "sentier des mégalithes", the Megalithic Trail in Erdeven which takes you along many megaliths. It starts at the Erdeven alignments*.

The weather did unfortunately not allow me to do that walk, so I have spent the last few days working on the new edition of my book and on improving my websites.

Finally today was somewhat milder and dry, so I decided to go back to the Carnac alignments*. I think it might be better for me to invest my time in making some videos of me dowsing these magnificent monuments rather than spending time on preparing to give a lecture in French, finding a venue for it and all the work that comes with it. It is very quiet in Carnac for the moment, hardly any visitors, just some locals. Ideal for making videos and enjoying the energies emanating from the Megaliths.

6. THE KERMARIO ALIGNMENTS

I decided to make videos at the sites that will be closing for the summer in a few days. You also don't know when they will close a site during the winter because I believe it's the town of Carnac that decides and owns the sheep that graze at the sites. The latter could be very well the case for the Kermario alignments* so I decided to make videos here first. I looked for the best spot to shoot a video of the alignments* and the magnificent Kermario dolmen*. I measured the energies first and found them to be representative for these

18

types of Megalithic monuments. This energy configuration is briefly explained at the beginning of this book.

Picture above and below: the Kermario alignments*

Picture: the Kermario dolmen

I then went to the tertre of Manio where I found that it is on a water source as I mention in my previous book. I measured, same as before, 12 signals for running water, only this time I found them with the negative energy signal which was completely neutralised last time I measured. Almost at the exact moment I was measuring here today, a super full moon was rising from the horizon and we are only a day from the spring equinox. Maybe this could influence the negative signal of the running water. The last time I was here there was a long drought, now several places were even flooded so there might be more running water which would implicate a stronger negative energy signal.

Picture: the alignments* and tertre of Manio

I also made a new discovery when I was checking the positive energies that are formed by the menhirs on top of the tertre. It seems, the tachyon energy* signal is not carried away in these streams into the land, the orgone* signal is though, as well as life force energy* and golden ratio* energy. Later research in Ireland does reveal that tachyon* can sometimes be carried away within the running streams. This might depend on the type of Megalithic Structure. I explain tachyon energy* elaborately in the glossary at the end of this book.

I then took a nice refreshing walk through the woods to the quadrilatère* and Giant of Manio and found both to show the same energies as last time. Also, two majestic monuments for my videos which I hope to start making tomorrow.

7. SPRING EQUINOX

Today is spring equinox but also a super full moon. It is a mild and sunny day. Ideal for making videos.

I made several videos at the Kerlescan alignments* in Carnac. I find it very calming and enjoyable to make these videos and just spending time amongst the stones. I also really enjoyed the cheerful chirping and singing of the birds.

Picture: the Kerlescan alignments*

I had some time left to make 2 videos at the Tertre of Manio. As I already measured yesterday, the signal for running water here is no longer completely transformed into beneficial energies.

Picture: the tertre of Manio

I did, however, discover something completely new today. When I tried to authenticate the signal, by changing the polarity of the magnetised rod in one of the arms of the antenna, I only found the positive signal for running water, not the negative signal!

So the Lecher antenna is only detecting a signal for running water in positive mode. Interesting!

In the evening I decided to go to the Kerzerho alignments* in Erdeven near Carnac because I wanted to make a video of the equinox sunset firing up the menhir in the form of a hand whilst measuring what is underneath. X marks the spot. A photo in color of this special standing stone at the equinox sunset is on the front cover of my previous book "Signpost to the Holy Grail...".

23

Picture: the standing stone at the Kerzerho alignments* which has the shape of a hand

I was well rewarded as the sun came from under the clouds just before it set and I got to make the most beautiful video for which I even got an audience because a bus of people had just arrived. They told me they were union reps on a tour for training newcomers. They seemed quite interested in my work. Some took a photo of the cover of my book promising they would give it some "word of mouth" publicity.

8. TERTRE, GIANT AND QUADRILATÈRE OF MANIO

Today is a magnificent beautiful day. Not a cloud in the sky. Early summer.

I set out to the Quadrilatère* and Giant of Manio to make videos and to see whether I can find out whether some energetic collaboration is in place between the two structures.

I did however make a stop first at the tertre of Manio again to check some reading I did here a few days ago, this being the one of the negative energy signal of running water. I was taught to put the magnetised rod in the left arm of the Lecher antenna* but I decided to put it in the right arm instead a while ago because someone told me that this is where a left-handed person should put it. My right hand is my receiving hand, with the magnetised rod in the antenna only the positive signal is measured so the person doing the measurements is protected from the negative signal. When the rod is inverted in the arm, the negative signal is detected instead of the positive signal. Once at the tertre of Manio I checked with the rod in the left arm and then in the right arm, to ascertain whether I would not be able to find a negative signal for the water current and this seemed the case but I am not certain because I was eager to get to the Giant and did not think of checking in positive mode. I am not really certain about which arm the magnetised rod should be put in to though, it could be that the entire Lecher antenna is magnetised when the rod is in contact with the antenna and that the arm in which the magnetised rod is put in to does not really matter.

It is a nice walk through the woods from the tertre to the Quadrilatère and Giant.

I know from my previous measurements of the Giant last year that it draws in 120 energy lines of the Hartmann grid*. With the new discovery I made at the tertre about the negative signal of the water current energy being neutralised, I wanted to know whether this might also be the case for the Hartmann energy lines, and guess what, very much so! Another interesting find!

This is also the case for the energy lines drawn in by the menhir of the Curry grid, the great global grid and the great diagonal grid*.

25

There are two water currents coming from the Giant which is an indication that it stands on a water source. Of the two water currents the signal for running water is completely neutralised (negative and positive signal), the currents can only be found with the signal for life force energy* into which the menhir has transformed the negative signals of the grids and the water signal.

I found the Hartmann grid* to be absent on the grounds next to where the grid lines are drawn to the menhir and only give a positive signal.

Just outside of the Quadrilatère* of Manio, however, the Hartmann and Curry energy grid lines can be found which were shifted out of the interior of the Quadrilatère* because a protective shield is in place, but these lines show a positive and a negative signal. Makes you wonder which structure would have been here first. Something to research further.

Picture: yours truly making videos at the Giant of Manio

26

Picture: Quadrilatère* of Manio

I made some lovely videos to promote my book and who knows, for an e-learning course about the energies of the different megalithic structures.

9. KERMARIO GIANTS AND DOLMEN

I then went to visit the Kermario giants and dolmen* again as I haven't done any elaborate readings here yet and it would be good to know how the energies relate between the dolmen* and the alignments* before making any videos.

The menhirs draw in 12 or 18 of the Hartmann grid* lines. The Hartmann grid is absent between the Kermario alignments* and the Kermario dolmen* but I do find 6 signals at the height of half visible fallen menhirs and even at certain which are completely in the ground. The Hartmann signals are as those of the Manio Giant, only for their positive signal. The ones that are half or almost completely buried are still fully functioning.

27

The dolmen* shifts out the Hartmann grid* from its interior, and here, like at the Quadrilatère* of Manio these Hartmann lines show a positive and a negative signal. For this to be the case, the dolmen* should, very likely, have been already in place before the Kermario Giant standing stones were put in place, otherwise these Hartmann lines would have been drawn in by the Giant standing stones and therefore absent just outside out of the confinement of the dolmen*.

I went to the "maison des mégalithes" where I got confirmation that the dolmen* are older structures than the alignments*. They don't know for sure what is the case of the Manio Quadrilatère*. Apparently, there is a theory that it once was used as a dolmen*, but also of it being a much younger enclosure*.

I met three people today who know what a Lecher antenna* is. Even someone who is sharing the house with me for a few days and who is also working with bio-energy. It is good that this wonderful precious instrument is starting to get the recognition it deserves.

10.LOCMARIAQUER –TUMULUS DE MANÉ ER HROECK

Today is another sunny day. The forecast for the next 10 days is nothing but mild sunny weather. I decided to go visit Locmariaquer. First stop was at the covered alley* of Les Pierres Plates. There is now a little fence around its perimeter and walking on or going into the monument is prohibited for security reasons.

It was really warm and very low tide because of "les grandes marées", the very low and high tides near the equinox full moon time of the year. Lots of people were collecting periwinkles, oysters and other delicacies the sea has to offer. I asked a local where would be best to collect them and I got some good tips. But, apparently the oysters are quite a hassle to cut loose from the stones they've attached themselves onto according to the person I spoke to.

I visited the tumulus* Mané er Hroeck. The tumulus* Mané er Hroeck or Faerie Stone or Sorceress Stone is one of the rare giant tumuli which are still here today. Its dimensions are 100 meters long, 10 meters high and 60 meters wide. No bones were found in this tumulus*. What was found inside are polished ax-heads in Alpine jade and pearls and pendeloques (drops) in variscite.

I am very fascinated by these polished Alpine jade ax-heads. This seems to be typical for the Morbihan Bay/Carnac region. It is thought that the rough jade was brought from the Alps to Carnac where it was polished, some for many hundreds of hours. These particular jade ax-heads, with the Carnac form, were also found in as far as Donegal in Ireland and also in Scotland.

The Mané Er Hroeck tumulus* is the oldest of its kind (prince like tumuli) dating back to 6000 years ago. The objects found in this tumulus* can be seen in the Museum of History and Archeology in Vannes which is only open during the summer months and which I went to visit later in the year when I was in Brittany again in September. It is very fascinating how these objects were meticulously put in this tumulus*. I wanted to know more so I researched and found information in a French book from 1891.

When the tumulus* was excavated in 1863, they found an engraved stone at the entrance which can be seen in the following drawing. Some believe it to be a depiction of the Mother Goddess in a crest, others believe it to be a cartouche with the "totem" of the chief.

Fig. 120. — *Totem* du Mané-er-H'oeck.

Picture: engraved entrance stone at tumulus* Mané er Hroeck

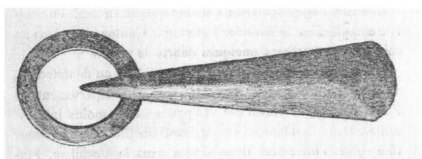

Fig. 121.— Hache et anneau du Mané-er-H'oeck, dans leur position originale.

Picture: original position in which the ring and ax-head were found, a unique feature

This is what I found in the French book about the excavations: "The Megalithic chamber shows another unique feature. On a layer of perfectly smoothed earth that has remained untouched since the construction was closed, rests a flat ring in green jade, slightly oval-shaped, on which rests the point of a magnificent ax-head in the same material. The jade ax-head is 35 centimeters long and is so well preserved that you get the impression it to come straight from the worker's hands. Behind the ax-head are two large drops in

30

calaïs, a kind of turquoise. Behind the drops is another ax-head in white jade and another drop. The ring, the large ax-head, the little ax-head and the drops were all placed in a straight line which perfectly coincides with the diagonal line of the chamber." The position of these objects can be seen in following drawing I found in the book of 1891.

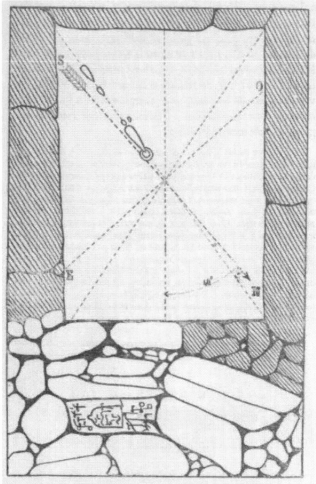

Fig. 122. — Aspect de la chambre du Mané-er-Hoeck.

Picture: original positions of the objects and engraved entrance stone found at the Mané er Hroeck tumulus* (S=south; N=north; E=West; O=east)

31

The fact that these objects are on the diagonal line is very interesting. Earlier geobiology research has revealed that whatever is on the middle lines of a room or dwelling, is amplified. These middle lines are also the diagonal lines of a room. I am tended to conclude that this was done for a reason and ask why jade and one in white jade. There are quite some most beautifully polished green jade ax-heads that were found at sites in this region, but white jade, that's very rare. Serpentine, green jade, variscite and white jade on the diagonal line in the chamber of this tumulus*... . Interesting to say the least!

The book goes on as follows: "Under this layer of earth, an irregular floor is found with traces of a wooden floor covering. This part consists of two distinct sections separated by a row of not very thick stones. In one of these sections, following objects were placed: 11 jade ax-heads, 90 in fibrolite, 3 drops in calaïs, 44 little pendant beads in quartz, agate and plain turquoise. Three cutting knives in silex and a large amount of carbon were also found."

Similar jade ax-heads were found at several other sites in the region too, even 4 were more recently discovered on the Quiberon peninsula in 2007 in the sod on the beach. They were found in two pairs with their points upwards and must have been there for thousands of years, untouched. This is very close to where the Kerbourgnec alignments* are, but these are now for the most part under sea-level and can only be seen at the very low equinox tides. These 4 green jade ax-heads are now part of the collection at the Prehistoric Museum in Carnac. I had the chance to admire them several times. They are most beautiful, although it might have been better not to take them away from their original position where they were so meticulously put in pairs. The Ancients must have had

32

their reasons to put them where they were…. A mystery yet to be revealed!

It is safe to go inside the tumulus* Mané er Hroeck, but you are asked not to walk on top of it so it won't become unstable.

Energy readings indicate the Hartmann and Curry grid* to be normal outside of the perimeter of the tumulus.

A protective shield is in place around the tumulus*: 13 Hartmann energy gridlines, 5 Curry energy gridlines, 1 great global and 1 great diagonal grid* line are shifted out to just outside the perimeter of the tumulus*, this at the side of the tumulus* where I did my measurements. The inside shows a signal for transformation, sacred, divine, tachyon*, life force energy* and orgone*. There is also a signal for cosmic and telluric energy. It is very difficult to walk around the tumulus and it encompasses a large area so I did not look for faults or water signals.

Picture: entrance of tumulus* Mané er Hroeck

I went to have a look inside the tumulus*. It goes very deep down a staircase.

I did not go in entirely and I don't think it is possible to stand in there. I just took a video to see whether I could see the same particles flying around as I did at the Pierres Plates during my visit there last year. I did see some particles but only very few.

I then went to the Locmariaquer megalithic visitor centre because I wanted to find out whether the energetic improvement I made to the "Grand Menhir Brisé" last year was keeping. The Grand Menhir Brisé or Tall Standing Stone is the largest standing stone in Europe, well was, as it is now broken in four pieces. More information can be found in my previous book about this monolith and the adjustments I made to it.

Picture: yours truly at the Grand Menhir Brisé - the Tall Broken Standing Stone

I was very happy to measure that it was now still active after my intervention and the signals seemed stabilised. The signal for

34

transformation coming from all 4 pieces now felt very strong and each part of the menhir was now drawing 12 energy lines of the Hartmann, Curry, great global and great diagonal grid*. The signals are, like at the Giant of Manio, only measurable with the antenna in positive mode. There is no signal in negative mode.

All 4 pieces give a signal for transformation, sacred, divine, life force* and tachyon energy* as well as golden ratio* and orgone*.

The fault line was also present same as at my last readings.

I then tried to see how far the Hartmann grid is neutralised on the ground and measured for approximately 25 meters. So this would mean 12 times 2 meters. Not 4 times (4 pieces of the menhir) 12 Hartmann lines times 2 meters (=width between two Hartmann lines). An interesting fact. Good to know.

I am very happy that the Grand Menhir Brisé has kept on functioning after my intervention as some say that it is part of certain "crystal grids"/geometrical or mathematical configurations with other Megaliths.

The Grand Menhir Brisé is made of orthogneiss which was brought here from at least 10 to 20 kilometers away, quite an undertaking this must have been at the time! And for what reason? Why here? And why orthogneiss?

11.ALIGNMENTS OF SAINTE BARBE, THE OLD MILL AND DOLMEN AND QUADRILATÈRE OF CRUCUNO

Today is Sunday. I spent most of the day answering students' enquiries and doing some other administrative tasks.

I finally had some time to go outside and enjoy the very fresh air and sun. The air is indeed very pure out here as witnessed by the amount of moss which grows on the trees. The more of this particular moss, the purer the air, similar to in Ireland.

I went to see to what is left of the Sainte Barbe alignments*.

Picture: the Sainte Barbe alignments*

And the alignments of the Vieux Moulin, the Old Mill. The Giants and the Menhirs of the Old Mill. Beautiful!

I could also see stones here that are almost completely buried. There must have been many rows of stones here during the Neolithic*.

36

Picture: Giant standing stones of the Old Mill

Picture: the little standing stones of the Old Mill

I took a lovely photo last year at the autumn equinox of the full moon just after sunset at the little menhirs of the Old Mill. I remember it well, a very special moment as I did not even know there were menhirs here at that point and I had just stopped to watch the full moon rise. The menhirs were a most pleasant surprise. The photo of this special moment last year is on the back cover of my previous book "Signpost to the Holy Grail….".

I then went to see the dolmen* of Crucuno and noticed that the roof slab is well over the orthostats (upright supporting stones) so I found this a good opportunity to check whether the roof-slab also plays a role in the formation of the protective shield that makes all negative energy shift out and indeed, it does. Good to know. It's a very rare chance to be able to verify this.

Picture: dolmen* of Crucuno

38

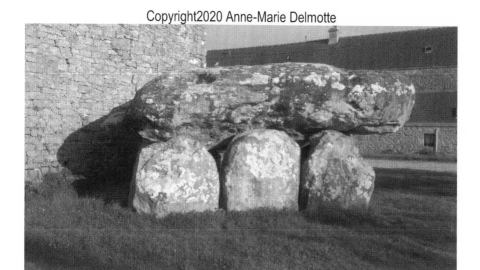

Picture of protruding roof-slab of dolmen* of Crucuno

I couldn't help but go visit the nearby majestic Quadrilatère* of Crucuno of which the corners are aligned to the summer and winter solstice sunrises and sunsets. Still as beautiful as ever and the energies are the same as I measured here last year. It's good to know that my measurements are reproducible.

Picture: quadrilatère of Crucuno

I'm now at "my" house in Plouharnel writing this. I hope to make some videos at the Kermario alignments* tomorrow of the Giants and the dolmen*.

12.   KERMARIO AND MENEC ALIGNMENTS

I managed to get to the Kermario alignments* part where the dolmen* and the largest of the Kermario menhirs are. It was very windy. I started with the dolmen*. As I did not get a chance to do any measurements of this dolmen*, I started by doing so. A shield against negative energies is in place and the orthostats and roof slabs all show a signal for transformation. I found a fault running perpendicularly under the dolmen*, but no signal for running water.

The negative energy of the fault is transformed into life force* and tachyon energy* and orgone*. The inside of the dolmen* also shows a signal for divine and sacred energy. I made a video of myself doing measurements at the Kermario dolmen*. I picked this dolmen* in particular because of its size and beauty.

Picture of the Kermario dolmen* and my Lecher antenna*

40

Then I recorded another video of measurements of two of the Giant menhirs. I had to do quite a few retakes because my tripod kept falling over from the wind. The weather seemed to shift from nice and sunny to overcast and windy.

Whilst I was making these videos I saw a man who came with a group on a bus carefully slip away from the group to do his duty against one of the menhirs. Really! I videotaped it! The manner of some people. No respect for these monuments!

After this I made good use of my time by looking for faults and water currents as I was not able to measure this in this part of the Carnac alignments* before.

I found 6 signals for faults, all running diagonally perpendicular to the direction of the dolmen. I also found three signals for running water, two parallel to the rows and one perpendicular to the rows but quickly showing a bend towards the other Kermario section that is beside the mill.

It might be important to mention that a pond very close to these alignments* was completely pumped dry some years ago which might have affected the running streams in this location.

Before heading back to my house I had a short visit of the end of the Menec site where I haven't been before and took some amazing photos of the menhirs reflecting in a pool of water still left from the rains they had here during the winter.

Picture: reflections at the Menec alignments*

13.   BROCÉLIANDE FOREST

I have been busy these last few days promoting my book in the Carnac region, and looking into booking a venue for a lecture I will be giving here next week.

I also returned to the Brocéliande Forest to do the same but firstly I wanted to visit a site I hadn't been able to see the last time I was here. It is at Tréhorenteuc on the way to Paimpont.

On my way there I saw a signpost with the mention "Allée Couverte" meaning covered alley* and I detoured to see this first. It is the covered alley* of Hino. It's on the edge of farmland and quite big and almost intact. I had a closer look and saw plenty of smaller white quartz stones. All of the orthostats and roof-slabs seemed to contain a lot of white quartz too or show veins of white quartz. One orthostat is in red shale. I wonder whether these white quartz containing stones were brought here especially and for what reason

42

they were put here. Is this covered alley* part of some white quartz crystal grid?

I checked the energies and these are the same as I found at other dolmen*. A protective shield is in place that shifts out the Hartmann and Curry gridlines of the earth*. These gridlines show a negative and a positive signal.

The interior shows a signal for sacred, divine, life force* and tachyon* as well as orgone*. A signal for transformation is coming from all the orthostats and roof slabs but also from the little quartz stones and the stones lying beside the covered alley*. I'm just wondering whether sending a certain energy to a stone will make it behave energetically. If this works, it could be something I could demonstrate at my lecture here keeping in mind that this might only work because there are a multitude of menhirs in the vicinity.

Picture: covered alley* of Hino

Picture: some of the white quartz pieces that can be seen at the covered alley* of Hino

I then went onto the site I set out to visit, this being the Val sans Retour or the Vale of no Return. The Brocéliande Forest is a magical place full of stories, myths and legends. The Val sans Retour consists of red shale and has a golden tree.

"It is the place of the legend about Morgane le Fée the Fairy, princess of Cornwall, half-sister of King Arthur and pupil of Merlin. Deceived by her lover Ginguemar, she traps him in a prison of the air with a powerful spell, along with all the Knights who are unfaithful in thought or deed who cross the threshold of Vale Perilous. Only a Knight of pure heart can release them... so the legend goes."

As I received the title of Knight in 2018 from our Belgian King it got me thinking; I am a person who is very much making an effort to discipline my thoughts and actions. I think I am quite safe to tread here. Not sure about releasing any unfaithful Knights though. Not my cup of tea ...

44

It is a lovely walk to the golden tree and the lake called "miroir aux Fées", the Fairy's Mirror. When there is a breeze and it's a sunny day, the sun is reflected all over the lake, magically beautiful.

After my lovely walk, a must do when in the region, I visited the church of the Grail. It is a little church with the most beautiful stained glass windows with depictions of the Holy Grail.

I then went to Paimpont to go to a shop I was at last year and where they were interested in my books and who I promised to show them once published but unfortunately the shop was closed, and so I went to another of which the owners are very friendly as well and very interested in my work. They asked me to come back to give a conference later in the year or next year. I even saw Morrigan I met last year, at the shop. What a small world, a coincidence? I was about to contact her to invite her to my lecture next week and "pouf", there she was. Well, the Brocéliande Forest is supposed to be a Magical Place. What a day!

14.   TUMULUS SAINT MICHEL – CARNAC BEACH – PROGRAMMING A STONE TO BECOME A STANDING STONE

Today I went to the Tumulus* Saint Michel and just sat in the sun, just as I did a few days ago. As it doesn't have any cosmic energy, it is a great place to get grounded which I need to get ready for my lecture (in French) and to promote my book which brings a lot of administrative work with it as well.

Picture: tumulus Saint Michel

I talked to a few people and the places for my lecture are almost all filled.

Things are also good because all the shops I introduced my book to, now have it on their shelves for sale. Some shops wanted to read the book first prior to stocking and told me they found it very good and interesting.

It is still the most beautiful sunny weather here and I took a walk on the Carnac beach looking for stones. I wanted to see whether I could make a full operating menhir or standing stone out of it. Something I could demonstrate at my lecture.

Picture: stone I found at Carnac beach

And you won't believe it, EUREKA, I did, with a simple gesture with the Lecher antenna* but I do think it is not that easy. I think my Reiki attunements and Crystal Reiki familiarisation sessions with certain crystals are somehow helping me with this. I think an attunement will be necessary to give someone else the ability to do so. Wow, I am making all these videos of the energies of the Megaliths to make a little e-learning course out of it but this seems to be the icing on the cake and a much bigger deal considering what all these beneficial energies produced by the menhir are good for.

I put the menhir in a little pot of earth and I am keeping it outside the house as I don't want to confine the orgone* which is produced by the menhir. Wow, it would be amazing if we could all have a working menhir in our garden. This might also have a positive influence on the weather according to some who made it rain in aried areas with the help of orgone* generators.

Picture: menhir in pot

15.   THE CARNAC ALIGNMENTS

Today is 31 March, it is the last day that the Carnac alignments* are open to the public. Tomorrow morning all the gates will close and the site can only be visited as a guided visit organised by the "maison des mégalithes" and this is until 30 September. It is sunny and very warm, a lovely day to do a last tour and make some last minute measurements.

I measured the last part of the Menec alignments* first. I was here before but had not looked for faults yet. There are two pools of stagnant water and this water has taken on the beneficial energies emanated and/or produced by the menhirs, this being sacred, divine, life force* and tachyon energy* and orgone*!

Picture: pool of stagnant water at Menec alignments*

I did not find a signal for running water. Two faults can be detected which run diagonally but almost parallel to the rows.

I then went to measure the Kermario site just behind the mill when coming from the "maison des mégalithes". I found no signal for running water but three signals though for faults. One perpendicular to the rows and two diagonal to the rows.

I also just enjoyed walking around the several sites which were open and had a closer look at most of the stones. Something I didn't have much time to do before, as I was always doing the measurements. A person might see faces on some of them.

49

Pictures: standing stones at the Kermario alignments*

16. LECTURE AND PROGRAMMING THE STONES

A few days ago I gave my lecture in French near Carnac on the different energy configurations of the megalithic sites.

As I wanted to film the conference, I was somewhat concerned that my wireless microphone would interfere with my readings with the Lecher antenna. I also wanted to make "sacred space" by means of 4 "Magical Stones" I brought over from Ireland but I could only check whether this would function on the day itself as the menhir I made from the stone I found on the Carnac beach had changed the energies at the house I am staying in and there was no way of testing this in the house. I would have to be at least 200 meters away from the menhir which is luckily the case for the venue where I gave my lecture.

The lecture was a success and I even made some new discoveries. As I made a rectangle/quadrilatère* energy configuration at the venue with the "Magical Stones", and then made a stone function as a menhir, I was able to check what happens to the gridlines when they have shifted out of the rectangle's protective shield. Readings indicate that these lines are not drawn in by the menhir but they too have been "fixed" by the protective shield created by the rectangular arrangement of the 4 "Magical Stones". These lines keep their negative energy signal when measuring with the antenna in negative.

I found the same result at the quadrilatère* and Giant of Manio site. Here also the gridlines are fixed by the protective shield.

Everyone wanted a functioning menhir to take with them and I did not disappoint my audience. We even witnessed that after programming a few stones, the new ones just started functioning on their own when I was measuring them before going over to the activation process! An indication that certain stones like to "mimic" behavior. An interesting notion!

There seems to be no end to my discoveries! I am so happy that I decided to write this book in order to share this knowledge and these discoveries with my readers, although I initially had no intention to write at all during this stay. I'm glad I did.

17.   ORGONE AND TACHYON

The day after my lecture I visited a "well-being" trade fair in the Brocéliande forest. A certain stall with very nice people was selling many orgonites and I asked them whether I could check them with my Lecher antenna* to which they agreed. I managed to measure 30 up until now but it's good to add to this number. I measured about 40 at the fair and they each gave a signal for the orgone* Lecher antenna* setting I discovered, none of them gave a signal for tachyon energy* though. I did measure tachyon energy* coming from orgonites from a friend who makes them in Northern Ireland.

A certain mineral needs to be added to the orgonite for the material to be tachyonised which is a patented secret. The lady from Northern Ireland must add this ingredient without knowing so.

18.DOLMEN OF MANÉ-KERIONED

Yesterday I visited a site which is lesser known near Carnac called the dolmen* of Mané -Kerioned. The site consists of a group of 3 dolmen*, surrounded by standing stones. Two dolmen* stand side by side, open to the south, whilst a third that is perpendicular to the two others appears to be an earlier construction. The site dates from 3500 before Christ.

Pictures: dolmen* of Mané Kerioned

Picture: standing stones at Mané-Kerioned and the ones across the road from the site
(indicated in the photo with 5 arrows)

When taking a closer look at the standing stones, and also the ones that are across the road from the site, they seem to be what's left of alignments* that consists of perpendicular rows.

One of the dolmen* is decorated with remarkable stone art of which some show this configuration of perpendicular lines (grids). Interesting! I will go back here later in the week to do some readings.

19. HOUSE CLEARING NEAR MENHIRS

Someone who was at my lecture asked me whether I would do a check of their house.

As it happens, this person has two fully functioning menhirs in their garden which makes working somewhat different from the usual geobiology house clearings. I found almost all geopathic stress* to be absent. What I did find was a lot of electromagnetic radiation and the middle lines amplifying a negative signal. Both very likely amplifying the negativity of the other as the house was only 6000 Bovis* units and the threshold for sick is 6500 Bovis* units meaning the house was not healthy. Something I have never done is to use my color filters on a dwelling but 4 colors did come up and after sending these to the house, the house went up to 23000 Bovis* units, meaning a place of spiritual quality. I did however still pick up the negative signal amplified by the middle lines so I asked the owner whether they wanted me to put in the energy configuration with my "Magical Stones" to which the person agreed. This shifted out the negative signal because now a shield against negative energy is in place around the house. The vibration was now at 26000 Bovis* units.

20.   MANÉ-KERIONED DOLMEN

After the house clearing I went back to the Mané-Kerioned dolmen* and alignments* to do readings. I checked the large dolmen* which is on the far left when looking from the road. It shifts out the energy

lines of the grids* which could be an indication that the dolmen*
was here before the alignments*, the protective shield thus fixing
these gridlines, as the menhirs of the alignments* can't seem to
draw these in. I checked the menhir, which is the one mostly to the
back next to this large dolmen* on its left, and it draws 12 lines of
each energy grid.

All three dolmen* at the Mané-Kerioned site were built on a
geological fault.

According to some, they are aligned to the winter solstice sunrise,
midday and sunset.

21.   MENHIRS OF MONTENEUF

I had an appointment in Paimpont today and just couldn't resist
making a detour to the menhirs of Monteneuf. I visited this site on
my way to Carnac last year and did not do any readings because I
wanted to research the Carnac alignments* first because the latter
have standing stones which were never redressed which is not the
case at Monteneuf, here all menhirs that are upright were
redressed. This was my chance to measure the Monteneuf
alignments*.

The site is just amazingly beautiful, even on a grey day like it was
today. The menhirs here are made of shale, a reddish/purplish like
shale. The name of this stone in French is schiste pourpre. The
colour gives the site its unique look.

Picture: the menhirs of Monteneuf

Readings indicate that the menhirs are fully functioning. The ones I measured draw in 12 energy lines of each grid and give a signal for transformation, life force energy*, sacred, divine and tachyon energy* and orgone*.

I found a signal for a geological fault running through the site. The fault gives one signal in positive and negative mode of the Lecher antenna*. The beneficial energies impregnate the fault except for the signal for transformation.

I noticed that a lot of the stones which were lying down showed veins of white quartz. I have been wondering about the red shale for a while and this seems to be an answer to my question. Shale is also known to contain mica, although it is not obvious when looking at the stones.

Picture: white veins in the red shale

I wanted to take a few very little stones with me to research but they were all functioning energetically as little menhirs. I decided not to take them as I wouldn't know what to do with them as I don't have a garden at my Belgian place and this was where I was heading after my stay in Brittany.

22.   LAST DAYS IN CARNAC REGION

These are my last days in Brittany for now. I am going to Belgium after this where I will prepare new e-learning courses and I wish to include my knowledge of the Megaliths and try to teach my students how to program the stones in order to make up for the Megaliths which were taken away over time and to benefit more people. In order to do so I have to make videos of me measuring them. After checking the ones I already filmed I saw that of a few, the Lecher antenna* signals were not clearly visible so I had to make new videos today. This was at the Giant and Quadrilatère* of Manio. Quite a challenge as it is the Easter holidays and many people are visiting the site. Some children just even left their bicycle against the Giant and took off. I wasn't aware that Megaliths are used as a bicycle stand. Luckily a nice family who regularly visit Carnac came along and kindly moved the bicycle for me so I could finish my last video.

I then had a lovely meal at the restaurant La Côte, which I know from my previous visit, across the road from the Kermario alignments*. They have a lovely garden where you can eat and the food is always exquisite. After my meal I had a long and very nice talk with the owners.

After this I made a video at the dolmen* of Crucuno where I found that there is a water signal running under it. I could not determine whether it is on a water source or whether it is a little stream running under it as one side of the dolmen* is not accessible because the dolmen* is right next to a house.

I then visited the Quadrilatère* of Crucuno where I checked the accessible outside perimeter for faults as there is a tree there which shows evasive growth. Measurement revealed no signal for faults or water but it stands next to a crossing of a Curry grid line and one of the great diagonal grid*. A Hartmann line runs also very close to where the tree is.

I am again very sad to leave this most beautiful region where you can find so many Megaliths, but I am already making plans to come back here to give more lectures, have a table at a trade fair, ….

I am however looking forward to going to Belgium too as I just got word that I will meet up with a Medical Doctor who has worked on about 10000 cases with the Lecher antenna*. After enquiring after his books which sadly are no longer available on the market, I was asked whether I would like to meet him. This is wonderful news, a doorway has opened to a treasure of almost forgotten knowledge.

GLOSSARY

Alignment: rows of standing stones. They can be combined, as at Carnac, with Megalithic enclosures and arranged in more or less parallel groups.

Bovis Scale: The intensity of the rays or vibrations, of a place, plant, food or object can be measured on the Bovis Scale in Bovis Units with a Lecher antenna. These are the units of vibrational quality or intensity of radiation of whatever is being measured. A healthy habitat and a healthy person should be at least 6500 Bovis Units. Values below 6500 Bovis Units indicate a qualitative energetic deficiency. Values above 6500 Bovis Units indicate a higher quality.

Cairn: heap or mound of stones covering a tomb

Covered alley: dolmen which has a long alley

Cromlech or enclosure: Megalithic monument made up of menhirs usually arranged in a circle or ellipse

Dolmen: Megalithic monument comprising a horizontal stone slab placed on upright stones. Dolmen were covered by a cairn or a tumulus when they were built.

Geopathic stress: geopathic stress is a form of trauma caused by disturbed or anomalous energies within the earth's mantle. The earth is surrounded by energy grids that can become harmful to all life. Geopathic stress has been implicated in a number of undesirable effects to human health, from simple conditions such as sleeplessness or confusion to highly dangerous ones such as cancer, decreased fertility, and auto-immune dysfunction. Dangerous causes of geopathic stress are the harmful underground water veins. Other causes of geopathic stress are Ley lines, global geomagnetic grid crossings, geological faults, underground caverns, and natural mineral concentrations that all exhibit similar effects. All these causes of geopathic stress can be measured with the Lecher antenna.

Golden ratio: The golden ratio is also called the golden mean or golden section, divine proportion, divine section, golden proportion, golden cut and golden number. It expresses a proportion with which a certain structure is build. This proportion (1.618...) can also be found in nature as in man and plant life. A Megalithic Structure, Standing Stone or building cut or built according to this proportion resonates with the energy of nature.

Grids of the earth: the earth is surrounded by several energy grids. These are found everywhere (in dwellings or on open ground) and are regular, although their mesh can sometimes be distorted by interference. Two grids are called after the person who discovered them. They go by the names Hartmann grid and Curry grid. The other grids are the great global grid and the great diagonal grid. All these gridlines emanate negative energy.

Lecher antenna: graduated scientific instrument able to detect and send energy. The instrument can be set to different frequencies in order to look for different energies.

Life force energy: combination of chi and prana which are both considered life force energies. They each have their own setting on the Lecher antenna.

Menhir: a Breton word meaning long stone. Megalith comprising of a tall upright stone or also known as standing stone. Menhirs can be isolated or linked to a group arranged in a line (=alignment), a circle (=stone circle) or a semicircle.

Neolithic: 9000-3300 BC (Before Christ): prehistoric period when man began to live in settled communities and grow crops and/or raise livestock.

Orgone: Wilhelm Reich identified a form of energy which he called Orgone and which is favourable to life, and is often equated with the "life-energy" in the human body. He found orgone would accumulate within a box whose walls consisted of alternate layers of metal and organic material (orgone accumulator or ORAC), and could then be detected by sensitive persons, or by a variety of physical means. Of these perhaps the best confirmed is the "temperature effect": the persistent small temperature difference between measurements taken inside an ORAC and a control box without metal. Much more recently, it has been claimed that by merely incorporating metal particles into a non-conductive matrix (usually synthetic resin) one can make a powerful source of orgone. Such material has been called "Orgonite", and is patented by Karl Welz. There are now a number of websites concerned with orgonite, some of which claim that it can be used to heal the environment. Another has photographs showing increased growth of plants close to orgonite. The author of this book, Anne-Marie Delmotte, has measured about 70 orgonites so far with the Lecher antenna and was able to determine a corresponding Lecher antenna setting for orgone.

On the other hand, orgone energy has also been shown to be able to push the body out of balance, making it very sick. It is a common fact that orgone energy can be transformed into a negative force field that can have disastrous effects on the body. Orgone changes into dead orgone when it is harnessed and/or unable to move freely. This is why I ask you to enter certain Megaliths which are very confined with caution. The beneficial energies are also present just outside of the structures which might be a better place to feel/take in these energies.

Quadrilatère: rectangular or square enclosure or cromlech of standing stones

Tachyon energy: The scientific community has shown that matter is nothing more than the condensation of a vibrating, universal substrate of subtle energy. This is the virtual condition that is known as zero-point-energy. Matter is created when zero-point-energy is transformed into tachyon energy. The tachyon energy is then transformed by the Subtle Organizing Energy Fields (SOEF'S) into matter and form.

The term SOEF (Subtle Organizing Energy Field) tries to describe how the descent of subtle energy in material form takes place and how it is organised.

Zero-point energy is present everywhere. Three major characteristics of zero-point energy:

First, zero-point energy is infinitely intelligent. Second, zero point energy carries within itself all possibilities that are necessary to create perfect forms. Third, zero point energy is formless and not manifest.

The first downward transformation from this formless non-manifested zero-point energy is the transition to tachyon energy. The prominent German physician researcher Dr. Hans Nieper described tachyon energy as a more or less condensed form of energy, the virtual condition on the way to change into a particle. The tachyon field exists on the boundary between energy and matter. Biologist Philips S. Callahan describes a tachyon as a "particle that moves faster than the speed of light". The energy keeps on transforming itself going downward and gets a form.

This tachyon model explains how the unlimited formless zero-point energy is condensing into tachyon energy. The tachyon energy is then transformed into specific frequencies through the Subtle Organizing Energy Fields. This energy is, according to Wagner and Cousens, transformed in the human body, in such a way that entropy is turned around. This has the effect that the ageing process is reversed. When a Subtle Organizing Energy Field is energised, it gets a better structure and organisation. This preserves the living organism, so that entropy, that is decay, is reversed, and also the ageing process.

Function of tachyon energy: slowing down the ageing process.

Tachyon energy has no frequency. It latently contains all frequencies in itself. Tachyon energy is the source of all frequencies. It enhances all frequencies without itself being a frequency.

Body and mind can only make contact with the zero-point energy through a tachyon field that is converted into frequencies by the SOEF'S. This is the key to recovery.

Tachyon energy energises all levels - the spiritual, the mental, the emotional and the physical - and balances them. Tachyon energy is capable of balancing both the left and right hemispheres of the brain. Research shows that tachyon energy has a strong influence on the brain. The SOEF'S convert tachyon energy in usable biological energy that is capable of waking or even reactivating slumbering parts of the brains. The physical body is the last part in the energetic continuum of man.

I was able to find a resonant setting on the Lecher antenna in order to measure tachyon energy.

To keep it somewhat easier to explain I mention several times how the Megaliths transform negative energy into tachyon energy. The presence of this tachyon energy has a more complicated explanation. The Megalith, and specifically the mica in the Megalith undergoes a tachyonisation, meaning that a self-perpetuating TAS (tachyon alignment* SOEF) is formed in which a high concentration of tachyons is found. Mica is a silicate and according to Wagner silicate is the easiest material which can be tachyonised and which forms the largest TAS of all known tachyonisable materials. As tachyon has no frequency but is the source of all frequencies, it is actually the TAS (tachyon alignment* SOEF) I measure with the Lecher antenna. So in fact the Megalithic Site draws on zero-point energy serving as a tachyonised material. The Megaliths serve as antennae, drawing and concentrating the tachyon energy out of the omnipresent zero-point energy.

Tumulus or barrow: a mound of earth and stones raised over a tomb. A cairn which is a mound of stones may originally have been a tumulus.

REFERENCES

-*The Lecher Antenna Adventures and Research in Geobiology and Bio-energy or Ascending the Veil on Secret Energies of Megalithic Sites, Energy Healing and Creating Sacred Space Using Free Magical Stones*, Dame Anne-Marie Delmotte, Delmotte Publishing, 2018.

-*Signpost to the Holy Grail? Infinite Energy with Infinite Possibilities! Dowsed with the Lecher Antenna in Carnac and Brittany France,* Dame Anne-Marie Delmotte, Delmotte Vibrating Energies Résonnantes, 2018.

-*Practical Guide for Dowsing with the Lecher Antenna – Elaborate Basic Training Course in Geobiology and Bio-energy*, Dame Anne-Marie Delmotte, Delmotte Publishing, 2018.

-*Document public - Carte géologique harmonisée du département de Morbihan- notice technique, BRGM/RP-56656-FR,* BRGM-Géosciences pour une terre durable, February 2009.

-*Wikipedia*

Carnac Alignements: https://en.wikipedia.org/wiki/Carnac_stones

Golden Ratio: https://en.wikipedia.org/wiki/Golden_ratio

Tumulus: https://en.wikipedia.org/wiki/Tumulus

-*Studies on "Life-Energy" by means of a Quantitative Dowsing Method. Comparison of orgonite with the orgone accumulator; spectrophotometric confirmation of its effect on water; nature of orgone* by Roger Taylor PhD; Syntropy 2012 (2): 17-32 ISSN 1825-7968.

-*The Mysteries-Unveiling the Knowledge of Subtle Energy in Ritual,* Bernard Heuvel, 2008.

-*Tachyon Energy – A New Paradigm in Holistic Healing* by David Wagner and Gabriel Cousens M.D., North Atlantic Books, 1999.

-*Points of Cosmic Energy* by Blanche Merz; The C.W. Daniel Company Ltd, 1987.

-*Nos origines-La Gaule avant les gaulois - d'après les monuments et les textes,* M. Alexandre Bertrand (membre de l'Institut), seconde édition entièrement remaniée – avec notes-annexes de MM. R. Collignon, Ernest Hamy, M. Berthelot, Ed. Piette et Salomon Reinach, Publisher Ernest Leroux Paris, 1891.

OTHER BOOKS/PRODUCTS BY THE AUTHOR:

-The Lecher Antenna Adventures and Research in Geobiology and Bio-energy or title (as e-book) Ascending the Veil on Secret Energies of Megalithic Sites, Healing Energy and Creating Sacred Space Using Free Magical Stones (English)

-Signpost to the Holy Grail? Infinite Energy with Infinite Possibilities! Dowsed with the Lecher antenna in Carnac and Brittany France (English)

-Unveiling Ancient at the Loughcrew Cairns – A Journey into the Discovery of the Subtle Energies – Measured with the Lecher Antenna (English)

-Practical Guide for Dowsing with the Lecher Antenna – Elaborate Basic Training Course in Geobiology and Bio-energy (English)

-e-learning courses: Dowsing with the Lecher antenna Bio-energy and Bio-energy Harmonic Remedies and Dowsing with the Lecher antenna Geobiology Basic Level and Geobiology Levels 1,2,3 and 4 (English)

-Guide pratique pour faire la radiesthésie avec l'antenne de Lecher - formation de base elaborée en géobiologie et bio-énergie (French)

Signe pour le Saint Graal? Energie infinie avec des possibilités infinies ! Mesurer avec l'antenne de Lecher à Carnac et en Bretagne France (French)

-cours e-learning: Apprendre à pratiquer la radiesthésie avec l'antenne de Lecher bio-énergie et Apprendre à pratiquer la radiesthésie avec l'antenne de Lecher géobiologie (French)

FOR FURTHER INFORMATION/SOCIAL MEDIA:

Website English: www.lecherantenna-antennedelecher.com

Email: annemarielecherenergy@gmail.com

Facebook Page (English): Lecher Antenna

Facebook Page (English): Megaliths and their special energies

Facebook group: Lecher antenna-dowsing-bioenergy-geobiology

Facebook Page (French): AntennedeLecher

Website French: www.antennedelecherradiesthesie.com

Facebook Page (Dutch/Flemish): Lecher antenne opleiding en boeken

Made in United States
Orlando, FL
15 August 2022

21082892R00040